MECHANICS·
MERCANTILE
LIBRARY.

Arthur F. Mathews '06

QUANTUM
LYRICS

Quantum Lyrics

· · · ·

P O E M S

A. VAN JORDAN

W. W. NORTON & COMPANY
New York · London

The epigraph on page 77 is from the poem "Sources" and reprinted with the
kind permission of Jay Wright.

For information about permission to reproduce selections from this book, write to
Permissions, W. W. Norton & Company, Inc., 500 Fifth Avenue, New York, NY 10110

For information about special discounts for bulk purchases, please contact
W. W. Norton Special Sales at specialsales@wwnorton.com or 800-233-4830

Manufacturing by Courier Westford
Book design by Helene Berinsky
Production manager: Julia Druskin

Library of Congress Cataloging-in-Publication Data

Jordan, A. Van.
    Quantum lyrics : poems / A. Van Jordan. — 1st. ed.
        p.cm
    Includes bibliographic references (p.).
    ISBN 978-0-393-06499-5
    1. Physics—Poetry. 1. Title
    PS3610.0654Q83 2007
    811'. 6—dc22

W. W. Norton & Company, Inc., 500 Fifth Avenue, New York, N.Y. 10110
www.wwnorton.com

W. W. Norton & Company Ltd., Castle House, 75/76 Wells Street, London W1T 3QT

1 2 3 4 5 6 7 8 9 0

For Aaron Jordan Jr.

January 23, 1928–October 22, 2005

Contents

$$\pi$$

Ω

· · · · · · · · · · · ·

"Time's arrow runs straight, but memory endows us with the capacity to bend that arrow into a loop, to revise the past in our minds to regain, even if in fantasy, that which was lost."
—ENDEL TULVING
Rotman Research Institute, Toronto

The Flash Reverses Time

DC Comics, November 1990, #44,
"Never Look Back, Flash,
Your Life Might Be Gaining On You"

When I'm running across the city
on the crowded streets
to home, when, in a blur,
the grass turns brown
beneath my feet, the asphalt
steams under every step
and the maple leaves sway
on the branches in my wake,
and the people look,
look in that bewildered way,
in my direction, I imagine
walking slowly into my past
among them at a pace
at which we can look one another in the eye
and begin to make changes in the future
from our memories of the past—
the bottom of a bottomless well,
you may think, but why not dream a little:
our past doesn't contradict our future;
they're swatches of the same fabric
stretching across our minds,
one image sewn into another,
like the relationship between a foot and a boot,

covariant in space and time—
one moves along with the other,
like an actor in a shadow play—
like a streak of scarlet light
across the skyline of your city
sweeping the debris, which is simply confetti,
candy wrappers, a can of soda,
all the experience of a day discarded
and now picked up
even down to the youthful screams of play
that put smiles on the faces of the adults
who hear remnants of their own voices
through a doorway leading back
to a sunrise they faintly remember.

Richard P. Feynman Lecture:
Intro to Symmetry

Love begins in the streets with vibration and ends behind closed
doors in jealousy. Creation and destruction. What do we pray
for but the equation that helps us make sense of what happens
in our daily lives? What do we believe in if not that which tells
us we're alive? Sex, laughter, sweat, and equations elegant
enough to figure on our fingers. Math is spirit and spirit is faith
in numbers; both take us to the edge but no further than we
can imagine. You don't believe in math? Try to figure the veloc-
ity of Earth's orbit around the Sun to land a man on the Moon
without it. You don't believe in God? Try to use math to calcu-
late what the eye does every second of any given moment. If Big
Blue tried to work that differential equation in our lifetime, it
couldn't. Mysteries inside mysteries in our own bodies of which
we can't make sense, another world waiting for a religion or cal-
culus to explain. Look into any mirror; it's like sitting in a the-
ater watching a silent movie, but you're the one pantomiming
your story. You think you have this world figured out, but you
can't tell which hand you're using and using and using. And
why do we try? You cannot solve for the use of one side of the
body over the other, so there is no single voice that emits from
it. You cannot solve for the harmonics of a dual body, facing
each other, both inquisitive. You cannot solve for the marriage
of opposites, their fit, their match, their endlessness. You cannot
solve for the morning stretch that calls to both sides, first this

one, then that one, aligning the day. You cannot solve for the
bass of one hand and the treble of the other, both keeping
rhythm hostage under the skin of the bongo. You cannot solve
for the balance of a locked door and a safe cracker's ear against
it and the move X number of clicks to the ~~left~~ and Y number of
clicks back past to the ~~right~~ and back past and back past till the
latch clicks open in your mind.

for Jesse Ingram

The First Law of Motion

Water damaged over time,
my 6th grade reader
cracks in my hands
and I smell the day thirty years before:

young male musk, after recess;
Paco Rabanne cologne;
Afro Sheen; and watermelon,
Now and Later candy.

"Teacher" describes the woman
at the front of the class.
Gerald and I laugh through her lesson,
slouched in our chairs in the back of the room.

She spots us. She tells Gerald, who can barely read,
to stand up and read a page aloud.
He shoots a glance at me for help.
For him, reading is an unfamiliar face.

And though I'm a strong reader, I'm too scared:
if I get caught, she'll walk me
into the hallway for swats
with her wooden paddle.

And when I hear his prayers
rising and falling from his lips—
each breath, each whisper,
carries no god's name, only mine—

I'm too afraid to speak.
He tries to laugh, but laughter
is another way to pray.
And he looks at me, again, to feed him words.

When I don't, his face folds
like a sheet of paper filled with mistakes.
Don't call it a grin stretching over her.
A smile shadows her voice.

"Sit down," she says.
But don't call this mercy.
Don't call this a lesson learned.
It's merely the question:

can he ever be again what he is right now,
a boy laughing with candy in his mouth?
Today, thumbing through pages
of my past—years too late

in a room cluttered with memory—
I try to feed words to an abandoned friend.

Black Light

Our bodies cast a shadow of one
body under a black-bulb pulse
in your mother's basement. Light, even

when it's black, moves faster than
youth or old age; it's the constant
in our lives. But I remember when

I thought your house—always ready
for a party, even during the week—
was the fastest element in my life.

Toenails, lint, teeth,
eyes—everything was holy
under the glow. I suspect

even my bones radiated
when we danced, which was always more
of a grind than a dance.

Whether the song sung came
from Rick James or Barry White,
we called what we did in the coatroom

dancing, too: My hands, ultraviolet
under your dress, but innocent. We
were only kids, after all,

I was 16 and you were a woman of 18.
Already, we knew how to answer each other
without asking questions, how to satisfy by seeing

what nearly satisfied looked like
in each other's face. This all before
I ran out to sneak back into my mother's

house in the middle of the night.
But, now, it's eight years later,
you're walking, it seems, so I offer

you a ride. And you look in and smile.
And when I see you I wonder
what would have happened

if we had stayed in touch. I have to get back
to work the next morning in DC,
a five-hour drive; it's near dark

and I want to get on the road before night
falls completely, but I stop anyway.
It's been too many years,

and I mistake your gesture.
Then I realize you
don't really recognize me,

until you back away and turn
on your heel.
Then a man with a Jheri curl

and a suit that looks like its woven
from fluorescent lighting
walks up and looks at me

like I wasn't born in this town,
and for the first time in my life,
I question it myself. He walks up as slow

and sure as any old player should on Sunday night.
While walking away, you two exchange
words. You don't look back. But

we see each other in our heads—aglow,
half naked—under our black-bulb pulse
in your mother's basement. Given a diadem

by the lucid night and the street lamp's
torch, the man wearing the fluorescent
suit casts a broad shadow

like a spotlight into which you step.
Maybe he's the reason we're here tonight
beneath these dim stars, casting

a light true enough . . . finally,
for us, after all these years, to see each other.

Orientation: Wittenberg University, 1983

A mock class. My mother and I
are the only faces of color.
I've never studied with white people,
but I've had my experiences.
And sometimes having experience
is the only way to study people.
But is there really a color
for ignorance when it hurts self? I

can see that I'm not ready.
The class is titled *The Fall*
and After, which is a study
of what happens after waking up: loss.
At this point, I understand loss
more than what comes before. A study
lending itself more to the act of falling,
an art in itself: How to appear ready

to step into the next stanza of life,
while tumbling down a page.
But, for now, in this classroom, the discussion
of Gauguin, Blake and the Bible
passes over our heads. Suddenly the Bible
is a foreign text, in the way they're discussing
it. I look down at the page
and it looks as blank as the life

I must have lived up to this moment.
The new students and the parents
get into this lecture, talking
about the art of falling. No hard times
or unfaithful lovers come up; this time
the blues can't frame the talk.
It's clear that, for some, life is a parent's
attempt to prepare a child for this moment:

when you walk into a room full of the educated
and you need to know what the hell
is going on. After class, my mother turns to me,
and says "You know, you don't . . ." she pauses,
"You don't have to do this," which gives me pause;
she's never said anything like this to me
before. I knew, even then, without skills any job is hell,
but then you prepare, you begin your education.

I decide what to do before she even gives me an out,
while we sit here, on the verge of knowing,
surrounded by people who sound like they know—
me with a Jheri curl, she in a wig—setting out

to make a mockery of class, my mother and I.

"Que Sera Sera"

In my car, driving through Black Mountain,
North Carolina, I listen to what
sounds like Doris Day shooting
heroin inside Sly Stone's throat.

One would think that she fights
to get out, but she wants to stay
free in this skin. *Fresh*,
The Family Stone's album,

came out in '73, but I didn't make sense
of it till '76, sixth grade for me,
the Bicentennial, I got my first kiss that year,
I beat up the class bully; I was the man.

But for now, in my head, it's only '73
and I'm a little boy again, listening
to Sly and his Family covering Doris's hit,
driving down I-40;

a cop pulls me over to ask why
I'm here, in his town, with my Yankee tags.
I let him ask a series of questions
about what kind of work I do,

what brings me to town—you know
the kind of questions that tell you
this has nothing to do with driving a car.
My hands want to ball into fists.

But, instead, I tell myself to write a letter
to the Chief of Police, to give him something
to laugh at over his morning paper,
as I try to recall the light in Doris Day's version

of "Que Sera Sera"—without the wail
troubling the notes in the duet
of Sly and Cynthia's voices.
Hemingway meant to define
courage by the nonchalance you exude
while taking cover within your flesh,
even at the risk of losing
what some would call a melody;
I call it the sound of home.
Like when a song gets so far out
on a solo you almost don't recognize it,
but then you get back to the hook, you suddenly

recognize the tune and before you know it,
you're putting your hands together; you're on your feet—
because you recognize a sound, like a light,
leading you back home to a color:

rust. You must remember
rust—not too red, not too orange—not fire or overnight
change, but a simmering-summer
change in which children play till they tire

and grown folks sit till they grow edgy
or neighborhood dogs bite those not from your neighborhood
and someone with some sense says Down, Boy,
or you hope someone has some sense

who's outside or who owns the dog and then the sky
turns rust and the streetlights buzz on
and someone's mother, must be yours, says
You see those streetlights on don't you,

and then everybody else's mother comes out and says
the same thing and the sky is rust so you know
you got about ten minutes before she comes back out
and embarrasses you in front of your friends;

ten minutes to get home before you eat and watch
the *Flip Wilson Show* or *Sanford and Son* and it's time for bed.
And it's rust you need to remember
when the cop asks, What kind of work you do?

It's rust you need to remember: the smell
of summer rain on the sidewalk
and the patina on wrought-iron railings on your front porch
with rust patches on them, and the smell

of fresh mowed grass and gasoline and sweat
of your childhood as he takes a step back
when you tell him you're a poet teaching
English down the road at the college,

when he takes a step back—
to assure you, now, that this has nothing to do with race,
but the rust of a community he believes
he keeps safe, and he says Have a Good One,

meaning day as he swaggers back to his car,
and the color of the day and the face behind sunglasses
and the hands on his hips you'll always remember
come back gunmetal gray

for the rest of this rusty afternoon.
So you roll up the window
and turn the music back on,
and try to remember the rust caught in Sly's throat—

when the song came out in '73,
although I didn't get it till '76,
sixth grade for me, the Bicentennial;
I got my first kiss that year.

I beat up the class bully.
I was the man.

Remembrance

Yesterday, today and tomorrow
string together a necklace of Wednesdays.
I go weeks like this:

where Friday nights spin
into a myth
I no longer believe,

where the words, *getting over*, italicize
into a mere idea.
I wake up some mornings,

and my grip feels so weak,
I check for stitches
at my wrists

to see if some young girl's hands
are now mine.
Some days my rough hands hurt

everything I touch;
some days I wish I had her
hands to make everything

I touch smile,
but, either way, I play
it off;

I get up; shower;
get dressed to report to work,
propped at my desk;

I speak—people look
like they hear a foreign language
all my own, a tongue

that no one finds exotic, simply strange—
and I'm talking about people
just talking;

I don't even broach the subject
of making love: there are no words
for this between our languages.

Some days I just want you to know
I remember, as a boy, walking home
from school, I saw Milton McKnight,

a kid we said was *a little slow*;
he was tied to a tree.
Three guys, for fun, were beating him

like a pedal on a bass drum,
but no music was coming out.
I want you to know, I remember

not Milton's blood but mine,
how I felt my blood coursing
through my body. This is how I learned

fear, how I had to tell my blood
to keep moving, relax. I did nothing.
I didn't want the three boys

to see me seeing them:
The kind of fear that keeps me walking
away from the scene, still.

Some days I can't even manage
a, *Good Morning*, or ask,
Man, did you see that game last night?

I can't do it.
The clichés bang against my teeth.
Some days I want to say something

to make you say nothing,
just look at me, deeply,
or for you to say something to me

so true it'll bring me to tears
years later. Some days my grip is so weak. . . .
Every day is a Wednesday. . . .

But, I digress: remember,
we were in the middle
of not talking

about love,
about how I open my mouth
and inside there's a small town

full of people who believe,
who actually believe
in Friday nights and even Saturday

mornings where men speak
softly and women walk slowly
and their memories hold

no threats for today.

The Green Lantern Unlocks the Secrets
of Black Body Theory

DC Showcase, September–October 1959, #22,
"Menace of the Runaway Missile!"

I did not flee to America;
I broke away from my planet
to save another: You see, I brought
the emerald wave of Oa with me
to Earth. I don't run to or away from the light;
I am the light. But there's also something
glowing within the darkness: more light
that simply can't break away and shine.
That's why I'm here. Consider
my home as the perfection of emitting
and receiving heat energy,
equally, in a body others
see as a mere void in space.

A view of my planet from Earth shows
a penumbra from one angle
and a full eclipse from the other, my origin
hidden like any immigrant's history
hides from the country of acceptance,
but, even here on Earth, my planet still radiates
within me underneath this mask.
I'll find a respectable man,
Hal Jordan, to carry my spirit,

and I'll do my best to fit in,
inking up the scene
to save it, an émigré from a planet
most Earthlings can't even pronounce.

A missile trained to attack
the White House in America,
as if 1,000 lanterns were shattered,
begs the question: should I get involved?
I have to answer:
whether I'm born here or just passing
through, life for me is as fragile
in the face of destruction
as when a gun is in my back
and I throw up my hands.
We're all equal under fear,
but we all huddle
under my lantern's emerald glow,
which powers my ring the way an oath
empowers the body speaking it aloud,
but, with that same force, squint your eyes
in pain: if you get too close to its truth,
it burns as much as it lights the way.
When I see the lantern
floating there, naked and cold,
I picture its potential power at rest,
what it takes in and gives back,
including light and the shadows
not allowing the light to escape.

π

•   •   •   •   •   •   •   •   •   •   •   •

"The social outlook of Americans . . . their sense of equality and human dignity is limited to men of white skins. The more I feel an American, the more this situation pains me. I can escape complicity in it only by speaking out."
> —ALBERT EINSTEIN addressing students at the historically black college Lincoln University, May 3, 1946

Quantum Lyrics Montage

FADE IN:

INTERIOR: ALBERT EINSTEIN's head, 1905—NIGHT

Thought Experiment #1: E = MC²

Tonight, somewhere lost in an occipital lobe
filled with geometry and seduction, I consider
Newton's laws of motion, force as matter accelerating
at the speed of sound and tearing through a room
like a man bent on destruction.
Someone releases the arrow from the bow
and then the heart feels the aim. Tonight,
I won't forget forces come in pairs;
two lovers kiss and someone gets hurt,
the action of love and the reaction
of disappointment are equal forces. Tonight,
even gravity will come into question;
I won't trust lifting my foot to walk
away for fear of no force strong enough
in desire to bring me back. Tonight,
I consider our bodies not simply as mass
but as energy—either potential, perched to attack,
or kinetic, incessantly wrapping arms around someone
in embrace—an energy, if you can imagine, traveling
not at the speed of light but squaring light's speed
opening into a blinding light that's more like a sun

emanating from a navel. Imagine
the century moving through its years
with the knowledge of our bodies
and with the knowledge of what our bodies
create and destroy, birthing and burying
our achievements and our mistakes.

FLASHBACK:

INTERIOR: Einstein's apartment, Switzerland, 1901—DAY

While on a temporary teaching position at a school in
Schaffhausen, Switzerland, a letter from MILEVA MARIĆ, who
will soon be his wife, arrives to Albert Einstein, held in his hands.

My dear, naughty little sweetheart,

I hate that you're not coming
to see me again! Unless you plan
to surprise me with a visit,

which would change the equation
inside me. I hope my sighs grow
deafening in your ears, thunder

expanding the air around you
as my voice. I hate
that you're quarreling still

with your parents, all for a little
human love from me. Infinite space
is so hard for people to hold

in their skulls, but they believe
in infinite happiness.
I don't understand this. It's much

harder to comprehend what men and women
share than the universe's infinity,
which is more difficult to grasp

lying in a bed alone. I miss you.
I search for a reason not to
long for you, some break

in our circle, but the ends continue
meeting. I have your shirt hanging
on my closet door to declare

it as the only flag to which I hold any
allegiance. I have a pair of your socks still;
they match but where's the symmetry

without your feet? I hate
to demand but if you value our intimate
rendezvous, you'll honor my privacy:

please don't share what I share with you
with anyone else. These letters
like me in this threadbare dress,

or a theory that puzzles but entices, whispers
like one of your angels in a tongue only for you.

CUT TO:

INSERT SHOT: Einstein's notebook, 1905—DAY

Einstein Defining Special Relativity

1: a theory that is based on two postulates (a) that the speed of light in all inertial frames is constant, independent of the source or observer. As in, the speed of light emitted from the truth is the same as that of a lie coming from the lamp of a face aglow with trust, and (b) the laws of physics are not changed in all inertial systems, which leads to the equivalence of mass and energy and of change in mass, dimension, and time; with increased velocity, space is compressed in the direction of the motion and time slows down. As when I look at Mileva, it's as if I've been in a spaceship traveling as close to the speed of light as possible, and when I return, years later, I'm younger than when I began the journey, but she's grown older, less patient. Even a small amount of mass can be converted into enormous amounts of energy: I'll whisper her name in her ear, and the blood flows like a mallet running across vibes. But another woman shoots me a flirting glance, and what was inseparable is now cleaved in two.

CUT TO:

INTERIOR: Home, Switzerland, 1908—DAY

*Mileva and Albert discuss their collaborations of 1905. While
Albert worked as a third-class patent clerk, Mileva worked as an
assistant in Heinrich Weber's lab, which allowed them to test the
"thought experiments" using Weber's equipment. Controversy
arises many years after both are dead, in the late 1980s, over the
three-page paper on Relativity in which the famous equation
$E = MC^2$ first appeared: in a Russian publication, both Marić and
Einstein's names appear; in subsequent printings, only Einstein's.
As his career ascended, he started spending less time at home.*

Collaboration
Mileva

Once, we slept under the night sky to understand
how the wind continues to make shadow
puppets of the trees. My body is a reed
instrument through which you breathe; a way
to make music, but it's not a song to which two
bodies can dance. Enough of this life as thought experiment.

Albert

Our love, despite the evidence, experiments
with the physics of simply being together. Understand
that a man must have an accomplice; two
hands are not enough: one flips the switch; shadows
hide the other. This is our world. Yet, there's no way
time will allow us to make love, test ideas and read.

Mileva

You find time to test ideas, travel without me and to read;
the unsolved problem is love. We are the experiment.
There's more than theories on my mind when you're away.
We have two boys who need a father who understands
their needs in the light of day, not just in the shadow
you cast at our door. You're one man, but your life splits in two.

Albert

Your eyes, black. Your hair, black. The door I enter and exit, too,
is black. Don't make the careless assumption of reading
my travels as anything other than necessity. I cast my shadows
where I must. You worked Weber's lab; I toiled experimenting
with boredom. Hiatuses are imposed, not chosen; understand,
at least, this much. Don't simply dismiss it away.

Mileva

Yes, yes, this "necessity" seems a way
for you to explain what you desire to
do, naturally. Trust me, my love, if I understand
the photoelectric effects of ultraviolet light on metal, I can read
a man. Have you forgotten? Your little urchin experimented
with the math, unveiling mysteries with you in the shadows.

Albert

I do remember the times we spent in the shadows,
my dear. You can trust these will never fade away
over time. Everything we do in life comes down to experiments
with love and curiosities. Lives should be experienced as two
children masquerading as adults. Although the public reads
the work of scientists and poets, this they don't understand.

Mileva

Albert, in shadows we find separate roads leading to our future:
a way to read the world's open-palmed offering,
an experiment to lead you to understand what you've lost.

CUT TO:

INTERIOR: Home, Switzerland, 1908—DAY

Albert Einstein confesses his infidelities to Mileva. This confession will ultimately lead to their divorce, but not for another eleven years. Their divorce agreement includes a stipulation that if or when he's awarded the Nobel Prize, he will give the prize money to Mileva, who maintains custody of their two boys, HANS ALBERT and EDUARD. Years later when he won the Prize, he honored this agreement.

Einstein

Your eyes hold enough lies
day to day, walking through the market,
like a woman living without
the sleight-of-hand of my skin, this peccadillo.

What rests inside me will rise
out my mouth to kiss you, to kiss
my confession into you each day, Mileva.
Will the truth offer you some freedom

or will I simply invite you to sit in my prison?
Wouldn't life play better if you visited me
in this cell, conjugal visits in which we pretend
we're free? I'll ask for you to come to me,

and I'll tell you to leave me, eternally
coming and going like the rain. I told you
only that I'm a man, only through how I hold you,
how I look in your eyes, like stars announce

they're stars by the dance and death of their light.
But you look sad as if you know now
what you must not know.

FLASHBACK:

INSERT SHOT: Albert replies to Mileva's letter announcing that she's pregnant. He addresses her by his nickname for her, Dollie. May, 1901—DAY

Dearest Dollie,

Exciting news, my dear—
I have been pacing elliptical orbits
around my desk, twirling
locks of my hair through thoughts,
and I may have an experiment for generating
cathode rays by ultraviolet light.
As for your news, promise me
you'll never cease being
the elegant equation of (little street urchin)
times (infinity) times the (speed of light).
Promise.

CUT TO:

INTERIOR: 1919—NIGHT

Mileva, recently divorced from Albert, is sitting at home peering into a photograph of Albert with his cousin and now second wife, ELSA, after their recent wedding.

Mileva

Possibly, she's more of a wife
but could she be more of a woman,
too? A floppy hat and a smile,
strands of hair astray,
a cloth purse and sound shoes
that led her to my husband who doubles
as her cousin—not much of a threat
but not so safe, either. He thinks
he'll experiment by taking a second wife.
Men behave as particles do
while being observed in light: they
respond differently in the dark
when you can't watch how they move.
I've worked math to prove to the world
that his thoughts were elegant; I've birthed
our children; I've laid my face in his hair.

I look at this couple and see myself in her.
I was young, scientific and willing
to fall into equations, through the infinite
yearning to understand physical laws

of action and reaction, of the force
of a mass in acceleration, of what's inert
and what work puts us back in motion.

People ask if I'm jealous
of science, as if I were a seamstress
instead of a physicist by training. Now,
another woman is by his side, and no
one asks my opinion. If they did,
I'd tell them the same truth I've said before:
what can you do? One gets the pearl,
the other the jewel box.

CUT TO:

INTERIOR: Berlin's Philharmonic Hall, 1921—NIGHT

German Nobel Prize–winning physicist and Nazi supporter
PHILIP LENARD holds a public forum of the Anti-relativity
League denouncing relativity and "Jewish physics"; Einstein
attends.

Einstein

And the sun will cease in a scene
much like this: ruined reputations remind us
that prominence doesn't protect people
who commit brilliance—the hubris to bear it, that is—

and defend friends against fools who hate
truth and tolerate nothing. Well, after this,
what can we believe in? How can we believe?
Philip Lenard makes moves

and trust turns from us to tax
the days when the sun says
this is a day that things must be done.
The revelry is reversed and heads reel

from the crimson morals credited to crowds
lusting after their own lives and livelihoods.
Look at him, festooned with the fear of friends.
A physicist serving anti-Semites,

booking Berlin's Philharmonic Hall, which,
even to refute relativity, reminds
us of what men must be mindful.
So I sit and listen to Lenard's followers:

anti-peace, anti-Einstein.
My seat in the audience warps space
and time; the entire hall turns
to energy, granting gravity to grow,

to stretch the light of their strained spirits,
frozen in time, in this temple of intolerance.
Of course, how these hovering black holes
do this—a prism of primal possibility—

is in my head, as I handle this hell.
Let them taunt; my mind is taut.

CUT TO:

EXTERIOR: Grand Canyon, 1931—DAY

*ELSA EINSTEIN poses with Albert for several photographs
taken at the Grand Canyon.*

Elsa

Little inspires awe in me
as much as Albert does,

but the Grand Canyon may hold
some of the secrets of the universe

and some of the grandeur of the man.
Here, I've donned Hopi headdresses

as they named him The Great Relative,
which carries more than a play on theory

in my mind. My cousin is my husband
and there exists no natural wonder

more perplexing than a woman who finds
herself in love with her best friend. America

has been good for his career and our marriage.
Here, the Americans from the Negro to the Hopi

seem to care for the man in my care.
He sees the citizens as a barefoot girl sees

a field of wild flowers in which she longs
to join. He says the Germans didn't know

what to do with him; at once, they saw him
as a stink flower and yet, with every new

honor bestowed on him by the world,
they kept putting him in their buttonhole.

CUT TO:

INTERIOR: "To American Negroes," *The Crisis*, October 29, 1931, office of Dr. W. E. B. DU BOIS—DAY

INSERT SHOT: Letter

Mr. W. E. Burghardt du Bois
Editor, *The Crisis*
69 Fifth Avenue
New York, NY

My Dear Sir:

There exists no erasure for race.
Not talking about it will not ease
the pain of questioning who is white,
Negro or Jewish, just to assess hierarchy
over humanity, hunger or hands reaching
for faces. The universe expands and the earth
orbits and we cannot change these phenomena
anymore than we should expect
to change someone's skin because
they're born closer to or farther from
the equator. Adding these factors won't equal
peace, unless we learn
they're pieces of a whole.

CUT TO:

INTERIOR: Theater, 1931—NIGHT

Premiere of City Lights *starring CHARLES CHAPLIN, New York City, Albert Einstein is Chaplin's invited guest. They sit together and the audience stands to applaud them.*

Einstein Ruminates on Relativity

Charlie Chaplin tells me
that the world loves him
because they understand him
and the world loves me

because they don't, which doesn't seem fair
but it's true: This is relativity.
Journalists ask for a definition,
but the answers are all around:

a woman loves you for a lifetime
and it feels like a day; she tells you
she's leaving, breaking it off,
and that day feels like a lifetime,

passing slowly. I listen to Armstrong
play his cornet and it sounds
like a Wednesday afternoon in heaven;
some hear Armstrong play

and it sounds like a Monday morning
in Manhattan. Some hear the war on the radio
and they hear acts of love; some
hear details of the war and it sounds

futile. Outside my window
people decry the rain;
somewhere else people pray
for rain to run down their faces.

CUT TO:

INSERT SHOT: Society page review of *City Lights*

A marionette without a master hand,
a mind playing between time
zones, a gesture of the shoulders—
one up, one down—

the gentleman tramp tips a top hat
instead of his derby and winks with both eyes
in the middle of conversation.
Einstein's mustache spreads

like smoke above his smile:
Two men with charisma.
A world begins with an embryo
of ideas, motion, velocity,

and ends with comic tragedy,
with thought experiments and proofs,
with flirtation, marriage and divorce.
Whatever we need to know

can be learned by watching a tramp,
a blind woman, and a park bench.
She hands him a flower, and the theory of everything
reflects in the misfit couple's eyes.

The world sees Chaplin and sees
the human condition; they see
Einstein and ask, Dr.,
where is your mind leading us?

CUT TO:

INTERIOR: Albert Einstein proposes a world government,
1932—DAY

Einstein on World Government

Sauntering into another world
war, to most, seems sensible.
This frightens me most of all.
Governments need the conscience of a scientist
like Alfred Nobel, who, after inventing
the most powerful explosive of his time,
gave birth to the most honored peace prize
of all time. We need to slay the quantum dragon
of atomic war, which can only be quelled
by dispersing its parts over the globe, sharing
the secrets of the bomb in its belly.

Some may say I think like a child
when it comes to politics. I say
world politics comes down to basic physics.
Since the ancient Greeks, man has known
the tools to measure movement.
To describe the movement of one body,
a second body is needed to refer to the first:
Earth's movement is compared to fixed stars;
a train's movement must reference the earth's surface.
So, like a child, I ask, if we blow up the world
to whom do we compare ourselves?

CUT TO:

INSERT SHOT: A scene from *The Battleship Potemkin* by
SERGEI EISENSTEIN.

Odessa Staircase

Men in uniforms carrying bayonets and guns
descend into the frame. Odessa citizens try
to ascend the stairs. A shot is fired from the soldiers.
A man with no legs is running
on his hands back down the staircase. A baby tumbles
down, step by step, in a perambulator. Cut to her mother's
eyes. Cut to the guns coming closer. The infant continues
down. The mother wedges through the bodies
with more rage than fear, looking up. Cut to the boots
of the soldiers stepping in pace. Cut back to the man
running on his hands. The sun in the face
of the people. Cut to a young boy who stumbles,
cut to the people pouring away from the march
of the soldiers. They run away from the sun.
Cut to the young boy being trampled
worse than to death, maimed. Down
the staircase, shadows descend; their guns,
in the foreground now, look bigger in the hands
of their shadows, than they did in the arms
of the soldiers casting them like a second unit of troops.
What happened to that baby? What happened
to that mother? All is drowned
by the silent flashes of fire from the barrels
of the guns, forever lighting the scene.

CUT TO:

INSERT SHOT: A scene from *Triumph of the Will* by
LENI RIEFENSTAHL.

The camera, installed on the flagpole, climbs up,
nearly three stories tall, to take
in the grandeur of German
seduction with rhetoric of peace
and the tension of swastikas and smiles.
See from the left they march in and cut across
while from the right the other regiment enters;
they make a sort of circle of uncertainty
around our lives. Listen, there's no
mention of race theory; there's not
one Jew in sight. Only black
boots, black and red flags and the chiaroscuro
of Nazi soldiers and their shadows,
of the people and the Führer.

CUT TO:

EXTERIOR: Princeton University, garden path, 1936—DAY

*Elsa Einstein, Albert's second wife and first cousin, has died
the day before.*

Einstein

This garden. This day:
opaque, fractal-plump clouds.
The air a jazz improvisation
of languages: strident, cool, chaotic

order. A spectrum of iridescent birds
play geometric games in the air
with magnolia blossoms, which follow
in their wake; I'm alone

with the voices seeping into the garden,
as if they, too, were planted here
from the other side of this fence of shrubs
surrounding me. I disturb the grass

with my sandaled foot and at times startle
some scavenging squirrel, cat or dragonfly
fleeing from me where perennials, irises,
pansies, or azalea sow scents

near moist with color from sunlight
and last night's rain and permeating
every movement, all matter doused
with fragrance singular to each genus,

singular to this space but collective,
a small country, insuperable
and striving through soil and danger,
as if all were dreamt by the same god.

CUT TO:

INTERIOR: Along with the influence of Einstein on
ERWIN SCHRÖDINGER'S discoveries on wave theory,
Schrödinger also shares Einstein's weakness for women. He
contemplates this at home with his wife, ANNEMARIE BERTEL,
1936—NIGHT

Erwin Schrödinger

Night, my wife, a star: an equation that makes me wonder
not about particles and wave patterns but why I was born.
I look at my wife, she presses her eyes along the sky; we're both
caught in a prism of starlight. Light asks no
questions; it carries answers; we understand light
teaches truth to the best of us—Newton, Einstein,

maybe even Schrödinger—even to the least of us. Einstein
once was considered the least, a man who wondered
if he'd ever make it, ever find his way to the light.
When he met his first wife, Mileva, a woman born
ahead of the time-space of the men around her; no
one would have thought Einstein could have both

brains and the good luck to marry a woman who was both
beautiful and a mathematician; Einstein,
who couldn't find a job, but who wouldn't take no
as a proof for his worth. Little wonder:
who knows what came before us until we search; we're born
with our eyes closed and screaming, screaming at the light.

And later we pray to the sun and the moon and the light
they reflect on us. My wife has enough light for both
me and the man that came before me. We're born
with no lovers on the horizon; we go along like Einstein
trying to find our purpose and then we look up and wonder
how we got here. We don't know

if we willed it, or if we fell in with the body next to us. No
single element in our universe can lead us to the light;
it's an equation of events we didn't see coming. We wonder
about the start to take the mystery out of the end. I ponder both
the women who came after my wife and the women Einstein
loved after his wife. Do we always need someone else born

to prepare us for the road ahead? As when an equation is born
years before an answer is known,
opening and closing constantly with meaning—as Einstein
discovered, finally, with the speed of light—
my mind, day after day, humbles with both
the result in elegance and the journey in wonder.

That love is born not out of deceit but from the quest for light,
makes me shake my head. No man has held wife and lover both,
not even Einstein, and not been changed, indelible with wonder.

CUT TO:

INSERT SHOT: Letter to President HARRY TRUMAN
from Albert Einstein supporting an anti-lynching bill,
1946—DAY

Trees need only to drop leaves to prove gravity.
The gravity of men hanged from trees is grave.
I speak against lynching; please, stand with me:
trees should only drop leaves to prove gravity.

To break the matrix of this enormity,
will take many voices to stave
off trees using bodies to prove gravity.
Mr. President, a man hanged from a tree is grave.

Trees should drop leaves to prove gravity.
Let's cut branches of men from trees;
there's a world watching this depravity,
and no laws of mere physics can end this tragedy.

The business of trees should end with leaves.

CUT TO:

INTERIOR: Meeting of National Committee to Oust
(Theodore) Bilbo. PAUL ROBESON speaks before Einstein,
November, 1946—DAY

Paul Robeson

In the back of my mouth sits
a song waiting to strike
combinations my hands can't handle.
If you ask my name

and hear my basso reply,
be ready for the truth. Listen,
I'm not feared for my arms or thighs,
not even for the song, but for the voice

they hear when I'm not panting
over yards, singing spirituals
or reading from a script. My voice
is as dangerous as any atom splitting

open, pouring its light
over a country or a belief.
Dr. Einstein, my main man, knows
the danger of having lunch with me,

of sitting with us, with the FBI
on his tail like German shepherds.
How revolutionary an act—
for saying, simply, what's complicated

about love and war. An elegant equation
can sum it up in a few factors,
but no one can do the math.

CUT TO:

INTERIOR: Princeton University, Cafeteria, 1949—DAY

*Princeton has recently admitted its first group of black students,
four in number.*

Princeton Cafeteria

If A plus B = B plus A,
A and B bear the ability to add up:
why isn't race always commutative?
Take this cafeteria, filled with fresh

minds, ready to embark on mastering
math, science, and the arts but the art
of conversation across race evades
these stars of society's future. Sure,

they're smart, but segregation and isolation
are cousins on this campus of coed high IQs.
When I ask, the whites tell me the Negroes
resist assimilation: they won't sit at our tables.

The Negroes tell me they're not wanted:
we're not invited to sit at the same table.
Princeton was Princeton before Negroes and will prevail,
as such, long after the Negro is a non sequitur to this southern

institution of the north. Innocent, you insist but,
this cafeteria is the feet of clay of this school.
Or, consider: A equals B when A and B are equals.
When one of the differentials is partial, how should

one solve it?

CUT TO:

INTERIOR: McCarter Theater, Princeton, NJ, 1955—NIGHT

The moment before MARIAN ANDERSON opens her mouth
to sing. Later that night, she will stay with Albert and his
stepdaughter MARGO.

Marian Anderson

In the audience, a woman blows smoke
through lipstick—holding
a cigarette in a cigarette
holder, in a gloved hand,

held to her mouth—
with more confidence than I blow
"Ave Maria" from mine.
Last night that woman slept,

secure, I'm sure, in her bed.
Or maybe she checked into the Nassau Inn,
sleeping in the arms of fine linen,
which made it easier for her to listen

than it was for me to sing,
not knowing where I would sleep
that night in Princeton, New Jersey.
Albert offered his home—

a haven in the midst of separate citizens,
of hotels and dress shops,
of diners and candy stores—at variance
with the theory of skin in Princeton,

which states that some of what allows me
to sing and stand erect through caterwauling
racists, might rub off on the denizens
if I get too close. I always stayed

in his home at 112 Mercer Street
whenever I dared to sing contralto
in that town, even when Princetonians
spat in my face, even now, years later,

after they've stopped.

FLASHBACK:

INTERIOR: Princeton Classroom, 1945—DAY

Einstein has read JOHN HERSEY's Hiroshima *in the* New Yorker.
He buys 1,000 copies to send to his friends around the world. Now,
he goes into the classroom to teach, facing the chalkboard to work
an equation.

Einstein Doing the Math

I turn to the black expanse of the chalk
board and the numbers spill
from my skull first and from fingertips
in time. Time in mathematics
brings complications, sequentially.
Numbers demand order and orders
demand numbers to behave. Otherwise,
one places one digit out of place
and an entire world loses
equilibrium. Someone determines that
one number is the temperature to freeze,
someone else realizes another number brings water
to a boil, but someone got the math wrong
and—now, if you'll allow me to dream—
the bombs pull us closer together
instead of separating the masses.
Working an equation is as tedious as a comedian
working a room, timing when to drop
the solution to our worries so profoundly we rear back
and laugh at them. Or, for those without

a sense of humor, math can be as simple as buttoning
a blouse, really: after you misfeed the first button,
though, every move of the hand, no matter how sincere,
becomes a lie.

CUT TO:

INTERIOR: Einstein's head—NIGHT

1:10 a.m., April 18, 1955, moments before his death in his sleep.

Thought Experiment #2: Toward a Unified Theory

My thoughts always arrive as potential energy.
It takes more than a lifetime to put them in motion.
Let's say, matter doesn't necessarily explode but opens.
Look inside each cell, each atom—quarks and quirks, too:
no animals in the testing, only light from bodies.
Let's say a fist comes toward your lips and you can't lean away
fast enough, because you're carrying that placard for peace.
It's not the mass of the fist that will kill you,
but the speed at which it comes
upon seeing your Jewish hair or black face.
Wavelengths absorb depending on their distance from the sun.
At sunset, we marvel at the colors and fall in love.
Look at the wavelengths in a stranger's eyes;
the closer you get, the lower the difference.
Get closer to the matter before you.
Place a bowling ball at the center of your bed.
The weight of the ball makes an impression.
Now, place a marble on the mattress.
Watch as it rolls toward the center.
When mass bends space-time, it's not gravity pulling us closer.
We simply fall into the dip
made around us. The day changes color,

as when someone walks through a door.
The night softens the songs,
hoping there's a way to absorb the voice.
Hoping there's a way.
Take all the phenomena—the skin, the sun, time,
the space of your bed of infidelity—
and although you come to realize
that thinking in public comes at great risk,
infinitely trying to unify all the loose strings,
you always believe it's the quest to understand the world—
even a piece of it, even after you fail—
that calls you to experiment with life,
but you pull yourself up suddenly, in the center
of the vortex, again, against judgment and advice,
all the unfathomable odds, realizing it's the struggle
to make the world understand you
that comes down to an equation that has no answer.

FADE OUT.

· · · · · · · · · · · · ·

Another god,
they say,
came like this . . .
    —JAY WRIGHT, "Sources (5)"

The Atom

DC Comics, June–July, 1962, Atom #1,
"Master of the Plant World"

It was as if no one had seen me

until I mastered the science

of shrinking my body

down to a particle

level, a basic element

of life so pure

that it was above

all frequencies of critique.

It was as if no one felt

my hand till it was pure vibration,

lighter than a gnat's wing,

on the back of their hand,

not till my touch became

a thought experiment

of memory: people wondering

if they could recall my skin

against theirs and if this recollection

was just that or a new experience—

something impossible yet flesh

and bone. I'm stronger than most men,

but I'm as weak as any man,

too. I fight to save the world

from destruction, but I also fight

to hear my lover, Jean Loring, say Yes

to marriage and to figure out my purpose

in this world. Do you think

if I could have managed my life

at 6' November, 1946 and 180 pounds

I would have shrunk

to near invisibility just to be seen?

Sometimes shrinking to the size

of a coin is a super power;

sometimes it's just a way

to find value in one's life.

The Superposition of the Atom

DC Comics, November 1963, Atom #9,
"The Atom's Phantom Double"

The tension of life is always death
and the twilight between the two

worlds. A phantom twin lives
inside. The day he comes out,

one of us must die.
Imagine a steel box with a cat

living in this four-cornered void
with a small vial of hydrocyanic acid,

an amount smaller than my eye,
and what will happen if it spills.

If the vial breaks, you see, it kills the cat.
The cat could live forever with this vial,

if I never look in the box,
or it could die quickly;

the vial could break within seconds.
But I never know if I never look

and the cat is forever dead
and alive. My phantom has existed for years

in limbo, believing life would be more
pastel if he were paying the bills,

sweating through rejection,
or figuring out what tie to wear

as Ray Palmer. I never know
if he's there or not, until jealousy

gets the better of him and he comes
out of paradox into a scene,

for which there is no future.
You can't blame him, though;

imagine being the cat, your life
determined by who looks inside the box.

Wouldn't you want to decide for yourself
whether you could be your own hero

or nemesis? Don't we all pray for the gaze
of some god who looks like us,

having mercy on what is seen?
How ridiculous my phantom looks trying to know

whether he's alive, standing translucent—
a mere shell of me on the inside—

trying desperately to look inside his own box.

for Chin Chong

The Atom and Hawkman Discuss Metaphysics

DC Comics, April–May 1969, #42,
"When Gods Make Madness"

A:

a sign in the sky

a Brahman poses as friend

a new enemy

H:

A Brahman without spirit? To defend

Against imposters of souls, to save others,

We must guard our own souls from this fiend.

A:

foes hide on borders

look behind you   nothing

in front   villains as brothers

H:

My planet exists far beyond, orbiting

Around a different sun where my wings

Seem common. Villains, friends the same—all vibrating.

A:

vibrations of kings

flutter of gnats both pests

for both   alarms ring

H:

How do we trust what we hear, and not test
What we see? Shadow and light both reflect
The same bodies, worlds of tension and rest.

A:

what do you detect
in my voice if not knowledge
of you and respect

H:

Sometimes I forget: comforts do privilege.
I question what I know and mysteries,
Too, which keeps me safe, but ruins my pledge.

A:

no fear   a series
of challenges helps the hand
to unmask faces

H:

Unveiling spirits takes a soulful plan.
Foes challenging the world is a constant
We can depend on.

The Uncertainty of the Atom

DC Comics, February–March, 1964, #11,
"Voyage to Beyond!"

When I move, I deceive
the eye of anyone looking,

shining the light on what
they approximate as truth in their eyes.

If I turn the dial on my belt,
I can shrink to the size of a particle

of light. I can pass
through a crack in a wall at a speed

specified, but who knows where I'll land,
once on the other side. Life

for me is a battle against villains
and self; I never know

how the day will be inked,
how the story will twist.

I might foil the plans of a foe
one day, and, later that night,

find him lurking over my shoulder,
a new episode. When a seed

is blown in the wind, you never
predict the landing, a rock

or a pasture could be in the plan.
Some days, I'm caught in the hands

of enemies and something as
common as turning on a light

while walking into a room, can change
the course of a day. I'll escape, maybe

weigh the villain down like a full-grown man
for a necktie or make myself

infinitesimal, less than a mosquito's proboscis,
which she only reveals, moments later,

by the bump on your thigh.
Or, in a breeze, I'll float like a leaf

you brush from your hair.
My life finds me with you whom I love

and you who will never love me;
of how you'll greet me, I'm never certain:

one minute I'm nowhere and the next,
sun streams through a crack in an open door,

and you cannot bear the weight of what you see.

The Atom Discovers String Theory

DC Comics, June–July 1964, #13,
"Weapon Watches of Chronos"

My plans were simply to escape
the grip of Chronos, my foe
who manipulates time and bends
it around me like a cage

with no space between the bars.
Shrinking to the size of no observed
size, shrinking to a particle with mass
but no weight, lighter than light,

was not the plan; but there I was:
so small I was lost between dimensions.
Space was merely a grid leading to more
grids. Our three dimensions were

doors leading to six more doors, opening
and closing with the force of a gale
through one and the force
of silent kisses through another,

oscillating faster than I could comprehend.
And if these dimensions were hidden
inside the crenulated folds of ours, I realized
I had disappeared. I thought

I was merely running away to come back
to catch my villain by surprise.
I thought I was tricking the trickster,
slipping his snare, but I simply ensnared

myself. I was lost in the dark, jumping
between lanes stretched before me
like violin strings. Every move
was vibrato—at first andante, then

andante un poco allegretto. Yes,
I was scared of the world within my own
in which even my tiny flashlight's beam
bent toward my boots in fear, it seemed,

as I moved along the spectrum,
plucking my way through the strands
of force and matter under my feet,
which were now, all at once, running closer

to and farther away from the answers
that would tie my world together.

Ω

· · · · · · · · · · · · ·

"Nothing changes that easily."

Fractals

The sun falls like a hemline breaking
over a man's shoe: I finally notice how
the horizon's line drapes in place,
mere months after my father's death.
I ask questions of him, again, about the future
of my journey: Will I sleep around
on my wife? Will I raise a fist
to her face and regret my hands
later? He tells me
We shall see what we shall see,
as I ponder the angle of his jawline
in this daydream. Zen was not his thing,
but when he didn't have an answer
he knew what to say. Nowadays, all images build
shapes around my life. An old movie airs
on TV, when I can't dream anymore.
These days, time doesn't really matter.
Two worlds meet constantly in my head:
Everything is love and loss.
I'm watching Glynn Turman as a young man
play his character, Preach, in *Cooley High.*
He's talking to Lawrence-Hilton Jacobs's character,
Cochise. Cochise is dead, though, and Preach stands
over his casket and starts: *Man, you know, sometimes*
I be walkin down the street or sittin around
And look up, and I expect to see you

comin around the corner, or hear your voice
callin my name or something.
His monologue once moved me,
but I don't believe it now,
while an old film reminds
me of what I've lost.
The falling sun is not enough;
there's only streamers of light—
not much, though—among the mountain of books,
CDs, receipts for all I didn't need,
and jewelry passed down from my father.
Outside my window, the sun keeps
falling along the sidewalk,
the buildings and the beaches, even as the night breaks
through; and all games, thefts, meals,
sex and dialogue fit
the shift in lighting, with the burden of my father's
memory heavy upon it. So frequently the dreams
volley from night to day—some thrown
on top of others, others flung
against the activities of the work week—
and the answers of my father, no wiser
to me now that he's dead, still
shape questions from my questions.

The Structure of Scientific Revolutions

I place the trombone back
on its stand, after attempting
"Summertime" in C major. Childhood
memories of band camp and lessons
stream to my embouchure, hands
and gut, after the ringing in my head
has passed, and the notes settle,
it seems, at my feet; I linger
on a photograph of my family: my parents,
brothers and nephew. I suppose
I stare off into it, longer than I mean to,
thinking of my father, dead eight months
now. The color image blurs
a bit on this print, not recalling his smile
as I recall it in memory.
And, also, the camera didn't know
this would be his last photo,
so the occasion was captured with people
moving like apparitions in the background
more than the spirit of the man in the midst
of it all. Even with digital cameras,
there's still a pointillist dotting flesh
on the faces of my family.
I shake my head to clear
the trance and turn on the television.
First, news of a woman raped;

the pundits keep the stress on "alleged."
I think of all the women who are
watching TV now. I turn the channel
to a rerun of *CSI*. All this technology,
yet we still can't prove rape. What's the use
of suspending disbelief? I turn the TV off
and sit back down to the horn,
but it's still next to the photo.
This time I notice my father's hand
at rest on my nephew's shoulder.
My nephew might be practicing his cello
as I sit in front of this brass,
which turns to folk art in my hands.
I don't really play anymore,
you know. I once thought music would be my life.
And it's simply too easy to try to play
and say, That's enough, too easy to say,
At least my nephew plays strings.
Nothing changes that easily.
It's in the way the mouthpiece
refuses to kiss me back, how the ceiling fan
whirs in the room, yet humidity
hangs in the air. A need builds in me
only after struggle builds around me,
a mythical ether challenging this horn
with its song stuck in its throat: memory,
the present moment and all the notes falling
between them, struggling to get out.

Sculpting the Head of Miles Davis

Secure the base
so the flesh will have something
to cling to. Wrap wire

around the wood
and fill with clay, liberally.
No, continue to add

clay—more than it seems
you will ever need for his
indented cheeks—and slap

more onto the base of the skull;
don't forget the constellation
of bones in the skull;

don't get hypnotized by the geometry
of the eyes; gouge your fingers
into his sockets—we'll deal with this later.

Pull back and follow the rhythm
of the jawline, rub your thumb
over the forehead; stab

your fingers into his cheekbones
raise them higher.
Doesn't his face

cast ribbons of shadow?
Doesn't he have cavernous
dimples? But don't make him

smile; imagine the teeth are behind
the sheet music of his lips;
imagine the tongue is aflame

behind the teeth; imagine
there's a voice scratching in the throat.
No. The temples sit too high;

the nose will not bespeak
his middle-class air;
raise the forehead,

straighten the nose bridge,
deepen the furrow of his brow.
Now, remember the look

in his eye back in '89 when
you saw him play at the Beacon Theater?
Can you see it yet? Stand back.

Tell me if the man whose face
you hold in your palms
could watch his mute drop from his horn,

at the *start* of his solo, pick it up—no
lowering of his head, no shrugging of his shoulders—
and go on to blow a phrase that still

trembles between your fingers.

for Raúl Acero

Richard P. Feynman Lecture:
Broken Symmetries

Symmetry walks between two worlds. To the hands it tries to touch us from either side; to the feet, it simply wants us not to stumble but to saunter; and, to the heart, it gives as much as it takes. Protons have neutrons; matter has antimatter. It's all a negotiation of will, a charade of dominance and submission, and we play like adults play with memories of our youth. We believe that love is equal to hate but nothing is perfectly symmetric. Instead, we should question why is the world nearly symmetric. Why, for example, does the earth orbit elliptically, as if these old hands had drawn the path, instead of following an elegant circle?

In the city of Nikko, Japan, stands the Yomei-mon gate. Elaborate in design, the gate has princes and lions and nymphs and other elements carved in—what appears to be, at least— perfect symmetry. But, if you look closely, you'll notice that one of the princes is carved upside down. And if you ask the people of Nikko *why*, they will tell you that it's carved so the gods won't get jealous of the perfection of man. But I put the mirror up to that statement and say that the laws of nature are nearly symmetrical because God didn't want to make man jealous of her hand.

And in the mirror, the clock ticks a little slower, the heart beats a little delayed. Watch the hand touch your face and, for a moment, one hand brushes both cheeks at once. But then you begin to pick the body apart: one foot is longer than the other, one breast hangs a little lower, one eye winks and the other can only blink and, suddenly, you're not the woman you thought you were. But then you look at a tree growing cherries or a flower sating a bee and you count the branches or the petals and you realize nothing is as beautiful as you once believed. And through our eyes, we continue coveting our reflections: The blade of grass wants to be a rapier; the clouds want to be smoke circles blown over lips; the eclipse wants to bring back the light.

R & B

Listen long enough to the radio, and you'll think
maybe C. Delores Tucker was right.
While the hip world is falling
in love with rappers with marquee-quality prison records,

I'm falling deeper under the spell of singers
who can still play piano. I never needed my female
vocalists to look good in a thong to feel their voices
in my bones; I never needed the male crooners to carry

guns to know they'd kill for love.
I said this the other night, driving
through Akron a week after my father's funeral,
trying to find a station without gangsta rap

or smooth jazz. For years, I watched
my father die, and when the day came,
my father had already predicted
the Chicago White Sox would win

it all this year. And on the TV
in his hospice care room,
as he took his last breath, Jermaine Dye hit
the first home run of the series. . . .

This is all over now. November.
Trees on front lawns rustle
and globes on streetlamps rattle
in a determined wind. I'm looking for fast food

and some music to change my mood
and the weather. In these silent moments
I listen, open to any sign or savior
I can find. I pass a half-lit, neon

Arby's Roast Beef sign
struggling through a dark night
of orbiting debris. I can barely tell
whether they're open or closed,

so I take a chance, turn around and fall
into its orbit as well. I walk in and see
two young black men—young brothers,
to me—with cornrows in their hair

and each carrying about 50 pounds
more than they should on their young bones,
standing behind the counter.
One brother asks me if he can take my order,

but I can't think about roast beef;
I think why is he so overweight as a teenager.
I think what is he going to look like when he's 40,
my age; I think what is he going to look like

if he makes it to 77, my father's age
when he died. I shake my head
and tell him I'm not ready; I try
to focus on food. I'm having a hard time,

because the music is so loud,
some alternative rock with an emo-boy voice.
Once I order, the brother who took my order starts
to argue with the one making my sandwich.

And when I think their argument is getting too loud,
when I think they're going to come to blows, they burst
into laughter and they stop and they look at me
and then they look at each other. The brother

on the grill nods to the brother at the register, who then,
with great earnestness, makes a request of me:
Sir, could you please tell him that that's not Al Green.
I say, of course that's not Al Green.

Then the brother on the sandwich says,
Naw, not what's playing out there—pointing toward me—
What's playing back here, pointing behind him.
The register brother says, *I don't think he can hear it.*

So they start to sing "For the Love of You,"
in harmony and falsetto. When they finish their set,
the brother on the register says, *Tell him that's Ron Isley.*
Yes, I exclaim, it *is* the Isley Brothers.

The sandwich brother confesses, *I know, I know* . . .
I wanted to see if he could tell the difference.
but you've got to admit, on some of those songs,
Al Green sounds a lot like Ron Isley.

Here I insert a caesura, while I ponder this cogent point:
You know, I say, you've got a point.
I never thought about it before, but it's true.
If you listen to Ron Isley, and didn't know the song,

one might mistake him for Al Green. At this moment,
I laugh with these brothers louder than I've laughed
since my father's death. I'm their last customer,
so the brother comes from behind the register

to let me out and lock the door.
I get outside and walk to my car
and get on with what's left of my life
as wind tears at the earth, leaves

settle around the streets,
and daylight subdues again,
till one day, through a near-closed window
in my mind, I'll see these two

young men among the generation
I once thought only sampled R & B;
I'll hear the vibrato in their voices,
and the mind's dust will wipe away:

with their song the sharp prongs of moonlight
will catch me smiling—a silly look of hope,
really; an equation of time and memory—breaking
through at least two generations of my blood.

for Bessie Jordan

Acknowledgments

Some of these poems, a few in slightly different versions or with different titles, have appeared in the following publications:

Black Renaissance / Renaissance Noire

Callaloo

Cave Wall

Ellipsis

Greensboro Review

Mipoesis

McSweeney's

New Engand Review

Redivider

Rivendell

Zone 3

Also, a different version of "Que Sera Sera" was printed in a limited edition chapbook under the same title from the University of Toledo's Aureole Press. Many thanks to Tim Geiger, Eric Elliott, Adam Schnell, and Adam Tavel for that wonderful program.

Thanks to Meta DuEwa Jones and Ellen Bryant Voigt for their keen, generous eyes on these poems; Jim Kakalios of the University of Minnesota, for his help with the physics in the Atom series; Angela von der Lippe, for her blessings on this project and, particularly, on the Richard Feynman poems; Linda Susan Jackson, E. Ethelbert Miller, Mitchell Jackson, Rodney Leonard, and Sebastian Matthews, for their support of this book; thanks also to Jim Schley, director of the Frost Place Poetry Festival, and Martha Rhodes, who got me that gig; thanks to Winfrida Mbewe and Alexander Cuadros for carrying the torch; and thanks to my family for keeping the faith in me.

Special thanks, more than I can say here, to Carol Houck Smith for making this book possible.

Notes

"The Flash Reverses Time." The Flash was introduced in the Golden Age of DC Comics and was revitalized in the Silver Age, after World War II. His super power is the ability to move at speeds nearing the speed of light. In the Golden Age, his true identity was Jay Garrick, and in the Silver Age and the present, Barry Allen, criminologist.

"Richard P. Feynman Lecture: Intro to Symmetry" and "Richard P. Feynman Lecture: Broken Symmetries." Dr. Feynman received the Nobel Prize in Physics in 1965 for the development of the space-time view of quantum electrodynamics. He is noted not only for his discoveries in physics, but also for his dynamic lecture style and personality. He was also a bongo player and a safecracker.

"Black Light." In the seventies and eighties, house parties were not complete without a black light bulb, usually purchased from a party-supply store. The bulb would make anything with a white hue to it glow: teeth, nails, eyes, lint on your clothes, etc.

"The Green Lantern Unlocks the Secrets of Black Body Theory." A Black Body is an ideal body or surface that takes in heat energy and gives heat energy equally. The Green Lantern is a complex DC Comics superhero who has many incarnations. Although a Golden Age superhero in the DC pantheon, he didn't become popular until the Silver Age. The most famous Green Lantern is Air Force pilot Hal Jordan, who is invited by the planet Oa to save Earth—and, consequently, the universe—by wielding the power of an emerald light conducted from a lantern to a power ring.

"My dear, naughty little sweetheart." Serbian mathematician Mileva Marić and Albert Einstein met in Zurich while both pursued Ph.D.s in physics. Although they divorced in 1919, when Einstein won the Nobel Prize in 1921, he gave the award money to Mileva for their children.

"Odessa Staircase." This scene is from the 1925 film *The Battleship Potemkin*, written and directed by Sergei Eisenstein. The scene, an early example of oppositional montage, has been structurally copied in many films including Brian De Palma's *The Untouchables*.

"*Triumph of the Will*." This propaganda film was made in 1934 by Leni Riefenstahl, under contract with the Nazi party. Considered a master-piece for its use of lighting, camera angles, and editing, it is in opposition to—and in dialogue with—the work of Eisenstein.

"Erwin Schrödinger." Erwin Schrödinger and Paul Adrien Maurice Dirac received the Nobel Prize in 1933 for their discoveries in wave mechanics. While vacationing in the Alps over Christmas and New Year's, 1925–26, with his mistress, Schrödinger returned with a wave equation for matter, which became the foundation for modern quantum theory.

"Anti-lynching bill." During the nineteenth and twentieth centuries, white mobs would hang minorities and immigrants for alleged and prosecuted crimes; often these crimes would be for perceived insubor-dination or false accusations of rape or attempted rape against white women. Despite the efforts of Albert Einstein, Asa Phillip Randolph, Paul Robeson, and numerous others, it was not until June 13, 2005, that the U.S. Senate formally apologized for failing to act on more than two hundred anti-lynching bills. To the outrage of many of their colleagues, over a dozen senators, including Mississippi senators Trent Lott and Thad Cochran, still refused to support the apology.

The salutation in the letter to "W. E. Burghardt du Bois" is as Einstein wrote it, "du Bois."

"The Atom." The Atom is a DC Comics superhero from the Golden Age. Although he was discontinued, he reemerged in the Silver Age with his own series. His superpower is the ability to shrink down to particle levels. His true identity is physicist Ray Palmer. His love interest is Jean Loring, a prominent attorney, who never agrees to marry him despite his proposals.

"The Superposition of the Atom." The superposition principle explains the addition of amplitudes of waves from interference. This occurs when an object simultaneously possesses two or more values for an observable quantity, often illustrated with a thought experiment known as Schrödinger's Cat.

"The Atom and Hawkman Discuss Metaphysics." Adventurer Carter Hall, also known as Hawkman, and the Atom discuss spirituality and their purpose as heroes after they are nearly killed in this classic DC Comic from 1969, "When Gods Make Madness." In this episode, they face an identity-changing, mind-controlling villain called the Brahman. To take on the identity of the Hawkman, Hall finds "Nth Metal" from the imaginary planet Thanagarian to fashion himself a pair of boots and a wing harnesses that would allow him to fly. The artificial wings give great speed and agility; the boots are indestructible.

"The Uncertainty of the Atom." The uncertainty principle relates to superposition. While for many years it was accepted that the only uncertainty in measurement was caused by the imprecision of one's measuring tool, it is now understood that no experiment or calculation can be accurate without accounting for its margin of error. Uncertainty explores the relative narrowness or broadness of the margin of error in a physical observation.

"Fractals." Geometric patterns repeated at scale to produce shapes and curves found in nature. A tree's twigs mimic the branches from which they grow, and, in concert with this, the veins in the leaves follow the shape of the twigs from which they grow. Fractals are widely used in computer modeling.

"The Structure of Scientific Revolutions." The poem takes its name from Thomas Kuhn's classic of the same title.

"R & B." C. Delores Tucker, the first black woman to serve as Pennsylvania's secretary of state, one of the founders of the National Political Congress of Black Women and a stalwart in the civil rights movement, is often noted today for her fight in the mid 1990s against misogyny and profanity in rap music lyrics and videos. She died in October 2005.

Selected Bibliography

Aczel, Amir. *Entanglement*. New York: Plume, 2003.

Broome, John, ed. *DC Archives: The Flash*. Vol. 2. New York: DC Comics, 2000.

Capra, Fritjof. *The Tao of Physics*. Boston: Shambala, 2000.

Einstein, Albert. *Ideas and Opinions*. New York: The Modern Library, 1994.

————, and Mileva Marić. *Albert Einstein/Mileva Marić: The Love Letters*. Princeton: Princeton University Press, 2000.

Feynman, Richard P. *Lectures on Physics*. New York: Addison-Wesley Publishers, 1965.

————. Lectures on Physics. Unabridged CD, volumes 1–8. Basic Books, 2006.

Fox, Gardner, ed. *DC Archives: The Atom*. Vols. 1–2. New York: DC Comics, 2003.

Gleick, James. *Isaac Newton*. New York: Vintage Books, 2003.

Gribbin, John and Mary. *Annus Mirabilis: 1905, Albert Einstein, and the Theory of Relativity*. New York: Chamberlain Bros., 2005.

Hawking, Stephen. *A Brief History of Time*. New York: Bantam Books, 1998.

Hersey, John. *Hiroshima*. Reprint edition. New York: Vintage, 1989.

Jerome, Fred and Rodger Taylor. *Einstein on Race and Racism*. New Brunswick: Rutgers University Press, 2005.

Kakalios, James. *The Physics of Superheroes*. New York: Gotham Books, 2005.

Kaku, Michio. *Einstein's Cosmos: How Albert Einstein's Vision Transformed Our Understanding of Space and Time.* New York: W. W. Norton & Company, 2004.

Laskey, Ronald C. "Time and the Twin Paradox." *Scientific American.* Special edition. June 26, 2006, pp. 20–23.

Milan, Popović, ed. *In Albert's Shadow: The Life and Letters of Mileva Marić, Einstein's First Wife.* Baltimore: Johns Hopkins University Press, 2005.

Newton, Isaac. *Principia.* Ed. Stephen Hawking. Philadelphia: Running Press Book Publishers, 2002.

Panek, Richard. "Relativity turns 100." *Astronomy.* February 2005, pp. 32–37.

Riedweg, Christoph. *Pythagoras: His Life, Teaching, and Influence.* Ithaca: Cornell University Press, 2002.

Smolin, Lee. "Atoms of Space and Time." *Scientific American.* Special Edition. June 26, 2006, pp. 82–87.

Trefil, James. "Relativity's Infinite Beauty." *Astronomy.* February 2005, pp. 46–53.

Tulving, Endel. "Episodic Memory: From Mind to Brain." *Annual Review of Psychology.* 53 (2002), pp. 1–25.

Veneziano, Gabriele. "The Myth of the Beginning of Time." *Scientific American.* Special Edition. June 26, 2006, pp. 72–81.

What the Bleep Do We Know!?. Dir. Elaine Hendrix. Lord of the Wind, 2004.

Wright, Robert. "Infidelity: Our Cheating Hearts." *Time.* August 15, 1994.